Terrific Trucks!

EMERGENCY TRUCKS

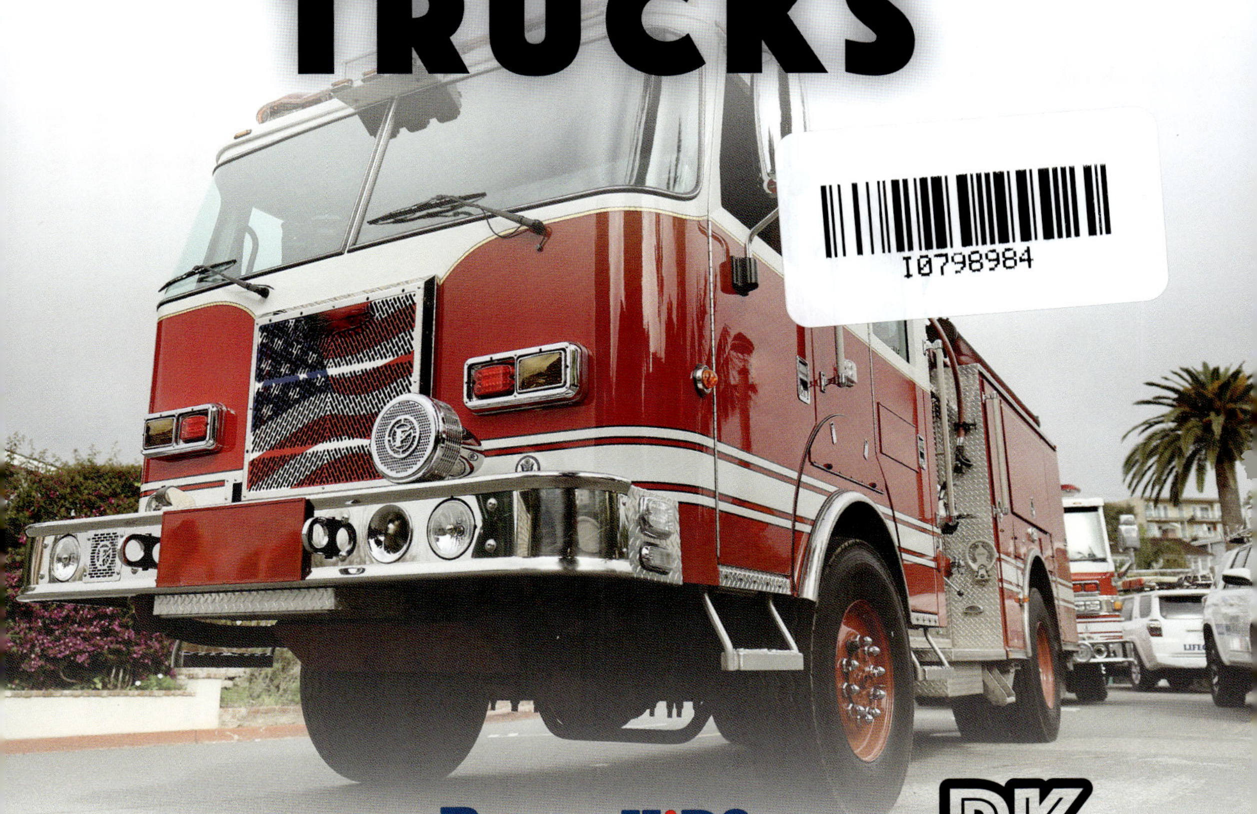

Ursula Pang

PowerKiDS press

PK Beginners

Oh no! A house fire!

Here comes
the fire truck!

FOGGY BOTTOM
23
ENGINE CO.
23

The siren is loud!

This fire truck has a ladder.

Hook & Ladder 148
It Ain't Easy
148
F.D.
NY
"It Ain't Easy"
148
FDNY

This fire truck shoots water.

ENGINE

Oh no! Someone is hurt!

Here comes
the ambulance!

2112
F.D. N.Y.
AMBULANCE
FDNY green
nyc.gov/fdny
United Church Of Deliverance
United Church Of Deliverance

Ambulance workers help.

A car is in the way.

The tow truck moves the car.

ZONA

Emergency trucks keep us safe!

Published in 2026 by The Rosen Publishing Group, Inc.
2544 Clinton Street, Buffalo, NY 14224

First Edition

Editor: Greg Roza
Book Design: Jeffrey Taylor

Photo Credits: Cover Heidi Patricola/Shutterstock.com; p. 3 sirtravelalot/Shutterstock.com; p. 5 Nicole Glass Photography/Shutterstock.com; p. 7 Mike Worley Photos/Shutterstock.com; p. 9 Kurka Geza Corey/Shutterstock.com; p. 11 Mike Brake/Shutterstock.com; p. 13 F01 PHOTO/Shutterstock.com; p. 15 Karolis Kavolelis/Shutterstock.com; p. 17 Riccardo Piccinini/Shutterstock.com; p. 19 Studio Peace/Shutterstock.com; p. 21 Gena Melendrez/Shutterstock.com; p. 23 AI generated image/Shutterstock.com.

Cataloging-in-Publication Data

Names: Pang, Ursula.
Title: Emergency trucks / Ursula Pang.
Description: Buffalo, New York : PowerKids Press, 2026. | Series: Terrific trucks! | Includes glossary and index.
Identifiers: ISBN 9781499451610 (pbk.) | ISBN 9781499451627 (library bound) | ISBN 9781499451634 (ebook)
Subjects: LCSH: Emergency vehicles--Juvenile literature. | Rescue work--Juvenile literature.
Classification: LCC TL235.8 P36 2026 | DDC 629.225--dc23

Manufactured in the United States of America